CLOUDS

JENNY MARKERT

CREATIVE EDUCATION

Designed by Rita Marshall
with the help of Thomas Lawton

© 1992 Creative Education, Inc.
123 South Broad Street,
Mankato, Minnesota 56001

Photography by Brian Aldrich,
Frederick Atwood, Gary Braasch,
Vera Bradshaw, Michele Burgess,
Greg Gawlowski, Mark Gibson,
Floyd Holdman, Image Bank,
Light Images, Ancil Nance, National
Center for Atmospheric Research/NSF,
Odyssey Productions, William Thauer,
Don & Pat Valenti, and Weatherstock

Library of Congress
Cataloging-in-Publication Data

Markert, Jenny, 1964–
Clouds / by Jenny Markert.
   p.   cm.
Summary: Text and photographs
examine the many types of clouds,
how they form, and the critical role
they play in climate, weather, and
the survival of life on Earth.
ISBN 0-88682-435-4
[1. Clouds—Juvenile literature.
2. Clouds.] I. Title.   91-8222
QC921.35.G46   1991   CIP
551.57'6—dc20   AC

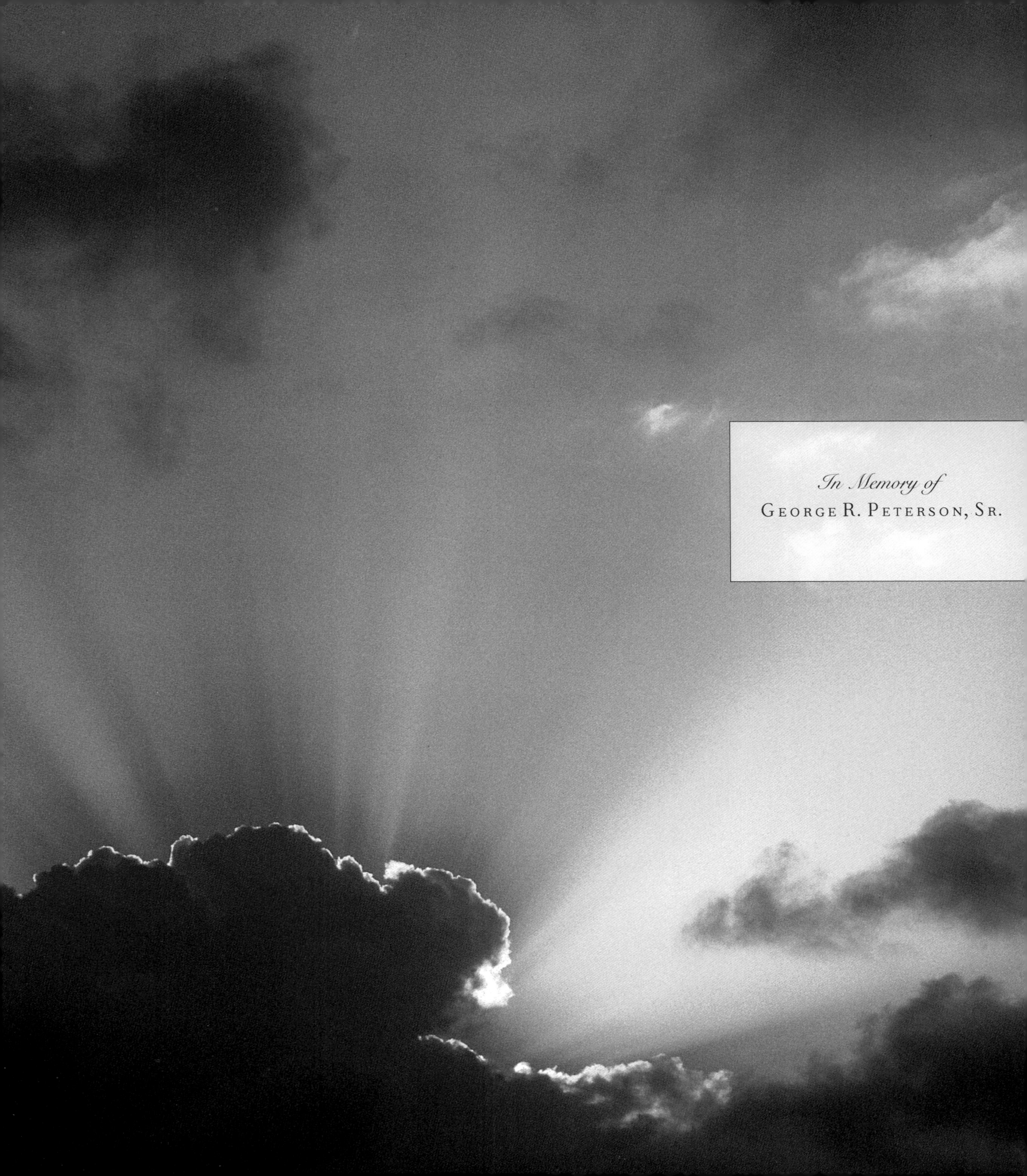

In Memory of
GEORGE R. PETERSON, SR.

7

People have always been fascinated by the skies. Although a ceiling of blue is usually a welcome sight, a covering of *Clouds* offers a much more interesting show. Clouds seem to appear out of thin air and sometimes disappear just as mysteriously. They can float lazily above the ground or cruise speedily across the sky. Some clouds pass overhead without releasing a single drop of rain, while others dump piles of snow or erupt with lightning and thunder. Although they are not always appreciated, clouds raise intriguing questions about the workings of nature.

Clouds can be very different.

9

A cloud is a collection of tiny water droplets, so small and light that they float through the air. Water enters the air through a process called *Evaporation*. Every day millions of gallons of water evaporate from the surfaces of plants and animals. An even greater amount evaporates from rivers, lakes, and oceans. When liquid water evaporates, it changes into a gas, called *Water Vapor*, and rises above the ground. However, water vapor is invisible, so it alone does not produce a cloud.

Water vapor rising.

On its journey through the atmosphere, the invisible water vapor naturally encounters changes in temperature. In particular, the higher the water vapor rises, the colder the temperature is. Cold temperatures have an interesting effect on water vapor. When water vapor is cooled, the invisible particles of gas join together, or condense, into tiny droplets of water. These water droplets are so small and light that they remain floating in the air. After countless water droplets form, they become thick enough to make up a cloud.

Vapor rises to join the clouds.

After a cloud has formed, its size and shape are constantly changing. One reason why clouds change in appearance is because air currents push clouds across the sky and distort their shapes. A cloud's appearance can also be affected by changes in temperature. If the temperature drops (and there is available water vapor), the water vapor condenses and the cloud grows bigger. If the temperature rises, the water droplets evaporate and the cloud shrinks.

❧

Although no two clouds ever look exactly the same, clouds do share similar characteristics. They often appear as thick sheets of gray, soft puffs of cotton, or wispy threads of silk. Scientists group clouds into three general classes: stratus, cumulus, and cirrus. The name they give each cloud reflects the cloud's general appearance.

A confused sky with many types of clouds.

Clouds that appear to form smooth, even layers are called *Stratus Clouds*. A layer of stratus clouds can dominate the sky, hiding the Sun and Moon behind a thick blanket of gray. Although stratus can form at any height, the bottom edge of most of them is usually less than 6500 feet above the ground. The lowest stratus cloud, known as *Fog*, hovers just above the ground.

Stratus clouds.

Unlike uniform layers of stratus clouds, *Cumulus Clouds* form billowing heaps that pile up into the atmosphere. Cumulus clouds are easily recognized, resembling gleaming white puffs of cotton with tops that look like cauliflower. With a little imagination, cumulous clouds can appear to be three-dimensional figures floating lazily across the sky.

Cumulus clouds.

Clouds that appear as thin, silky webs or wispy streaks are called *Cirrus Clouds*. Cirrus clouds form at least four miles above the ground, with some reaching altitudes of over ten miles. At these heights, temperatures are far below freezing, even on hot summer days. Because of the cold temperatures, cirrus clouds consist of tiny ice crystals, rather than droplets of water. Some cirrus clouds travel over two hundred miles per hour; however, they usually appear to be standing still because they are so far above the ground.

Cirrus clouds.

In addition to describing the three main classes of clouds, the terms stratus, cumulus, and cirrus are often combined to describe many other interesting types of clouds. For instance, *Stratocumulus Clouds,* as their name suggests, are layers of heaping clouds. *Cirrocumulus Clouds* are wispy, white puffs that resemble the scales of a fish. Another type of cloud is the *Cirrostratus,* which forms a thin, silky layer over most of the sky. Often, a rainbow-colored halo shimmers around the Sun or Moon as their light passes through the thin cirrostratus.

Page 18: Cirrostratus clouds with a halo.
Pages 18-19: Stratocumulus clouds.

Regardless of its name, from the thick stratus cloud to the high-flying cirrus, every cloud contains some form of water. However, all clouds do not release precipitation in the form of rain or snow. This is because the water droplets in a cloud are not heavy enough to fall to the ground. The water droplets simply float through the air, blown about by air currents.

In order for a cloud to release precipitation, the tiny droplets in the cloud must combine to form a larger, heavier drop of water. In fact, it takes more than a million cloud droplets to make a single drop of rain. Usually, cloud droplets do not combine unless the cloud contains tiny particles called *Ice Nuclei*. Ice nuclei give the drops of water something to collect on to form a larger drop of water. Most commonly, ice nuclei are tiny bits of salt, soil, or dust that float through the atmosphere.

Crashing surf throws off ice nuclei.

23

Sometimes natural ice nuclei are not present in the air. Without these tiny bits of debris, the water droplets in a cloud cannot combine to form a raindrop. If rain is badly needed by farmers, or the ski slopes are lacking snow, people can supply the ice nuclei that are necessary for precipitation to form. In a process called *Cloud Seeding,* airplanes fly into clouds and drop crystals of solid carbon dioxide (dry ice), which serve as ice nuclei. Cloud seeding can also be done from the ground, by releasing silver iodide smoke that rises into a cloud.

Dust storms send particles into the clouds.

25

Whether they are supplied naturally or artificially, ice nuclei are blown about in the cloud and collide with the floating droplets of water. When ice nuclei touch the water droplets they "trigger" freezing of the liquid. As more and more water droplets freeze onto them, the ice nuclei grow into larger crystals of ice. Eventually these bulky ice crystals become too heavy to continue floating in the air, and they drop out of the cloud as precipitation.

Ice crystals in a cloud.

27

Although most precipitation begins as tiny crystals of ice, the type of precipitation that reaches the ground depends on the temperature of the air. If the temperature is below freezing all the way to the ground, the crystals remain frozen and fall as flakes of snow. When the air temperature is slightly above freezing, some snowflakes melt into rain. This mix of snow and rain is called *Sleet*.

Snowfall in Idaho.

When the air temperature is well above freezing, all the ice crystals melt after dropping from a cloud. However, the resulting drops of water do not always reach the ground. The smallest drops of water evaporate on their journey downward, producing *Fall Streaks* that appear under the cloud. Raindrops that are slightly larger usually survive evaporation, but may take over an hour to reach the ground because they are so light. This precipitation is known as *Drizzle*. Only the largest drops of water reach the ground as a heavy soaking rain.

A rainstorm approaches.

Hail is another fascinating product that falls from clouds. Like snow and rain, hail forms from crystals of ice which are tossed about in a cloud. However, unlike other types of precipitation, hail only forms when there is a strong updrafting air current. These violent winds keep the ice crystals aloft after they would normally fall to the ground. The growing clumps of ice may toss about in the cloud for over an hour, until they become too heavy for even the strong air current to keep them aloft. Then, the chunks of ice drop out of the cloud, usually in brief, heavy showers called *Hailstorms*.

Hail-covered landscape.

Most hailstones that reach the ground are about the size of peas. These bits of ice seem to dance on the grass and pavement after their long fall from the sky. Much more threatening to life and property are hailstones that grow to the size of walnuts or even tennis balls. Hail this size can strip leaves off trees, flatten crops, smash windows, and even dent cars. The largest hailstone ever reported was about the size and weight of a melon!

Hailstorms are not very common because unusually strong updrafts are needed for hailstones to form. Generally, hail forms in only one type of cloud—the towering, dark rain clouds called *Cumulonimbus* (nimbo meaning rain). Besides releasing hail, cumulonimbus clouds often produce lightning, thunder, and torrents of rain. Because of the loud thunderstorms they produce, cumulonimbus clouds are commonly called *Thunderheads.*

Pages 32-33: Severe storm cell with cumulonimbus cloud.

Few things are as exciting as a *Thunder-storm,* with the ominous gray of the thunderheads, the bright flashes of lightning, and the piercing cracks of thunder. Before a bolt of lightning streaks from a thunderhead, hot air, cold air, water droplets, and ice crystals toss and turn within the cloud. The friction caused by this churning generates electricity. This process is similar to the way you build up electricity by rubbing your shoes or socks on carpet. However, the amount of energy you generate is very small, and is released as a small spark, whereas the churning in a thunderhead produces enormous amounts of electrical energy. When the energy is released, a brilliant flash of lightning streaks through the cloud, and sometimes, all the way to the ground.

Lightning in the clouds.

Always accompanying a bolt of lightning is its companion, the familiar crack of *Thunder*. Thunder happens when lightning tears a path through the air. As it passes, the streaking bolt of lightning heats the surrounding air to more than 60,000° Fahrenheit. The heated air quickly expands and then contracts, producing a loud, startling clap of thunder.

Lightning strikes a road.

Anyone who has been in a thunderstorm knows that you always see a streak of lightning before you hear a crack of thunder. Lightning is seen before thunder is heard because light travels faster than sound. Light travels so fast that you see a bolt of lightning at nearly the same moment it strikes, no matter how far away it is. On the other hand, it takes some time for the sound of thunder to travel to your location. By counting how many seconds pass between *Lightning* bolt and thunder clap, you can determine how far you are from the storm. Depending on wind conditions, every five seconds between the bolt of lightning and the clap of thunder equals approximately one mile.

Intense lightning and thunderstorm.

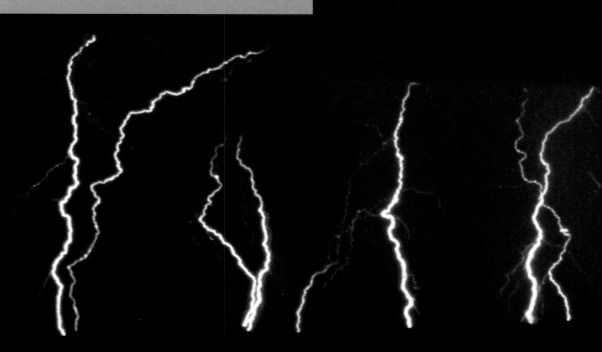

For some, a day of cloudy skies and rain-showers can spoil an afternoon picnic, while for others, such as farmers, it can answer a prayer. Whether we are elated or depressed by them, *Clouds* provide precipitation that is needed by all forms of life on our planet. But for humankind, clouds offer more than the materials for survival. Clouds spark our imagination with their shapes, inspire our creativity with their beauty, and challenge our intellect with their behavior. Whether we look at and admire them, or tremble and hide from them, clouds will always be a source of fascination.

Clouds viewed from an airplane.